The Literature of Cinema

ADVISORY EDITOR: **MARTIN S. DWORKIN**

INSTITUTE OF PHILOSOPHY AND POLITICS OF EDUCATION
TEACHER'S COLLEGE, COLUMBIA UNIVERSITY

THE LITERATURE OF CINEMA presents a comprehensive selection from the multitude of writings about cinema, rediscovering materials on its origins, history, theoretical principles and techniques, aesthetics, economics, and effects on societies and individuals. Included are works of inherent, lasting merit and others of primarily historical significance. These provide essential resources for serious study and critical enjoyment of the "magic shadows" that became one of the decisive cultural forces of modern times.

History of the Kinetograph, Kinetoscope & Kinetophonograph

W. K. L. Dickson and Antonia Dickson

ARNO PRESS & THE NEW YORK TIMES
New York • 1970

Reprint Edition 1970 by Arno Press Inc.
Reprinted from a copy in The Museum of Modern Art Library
Library of Congress Catalog Card Number 79-124005
ISBN O-405-01611-5
ISBN for complete set: 0-405-01600-X
Manufactured in the United States of America

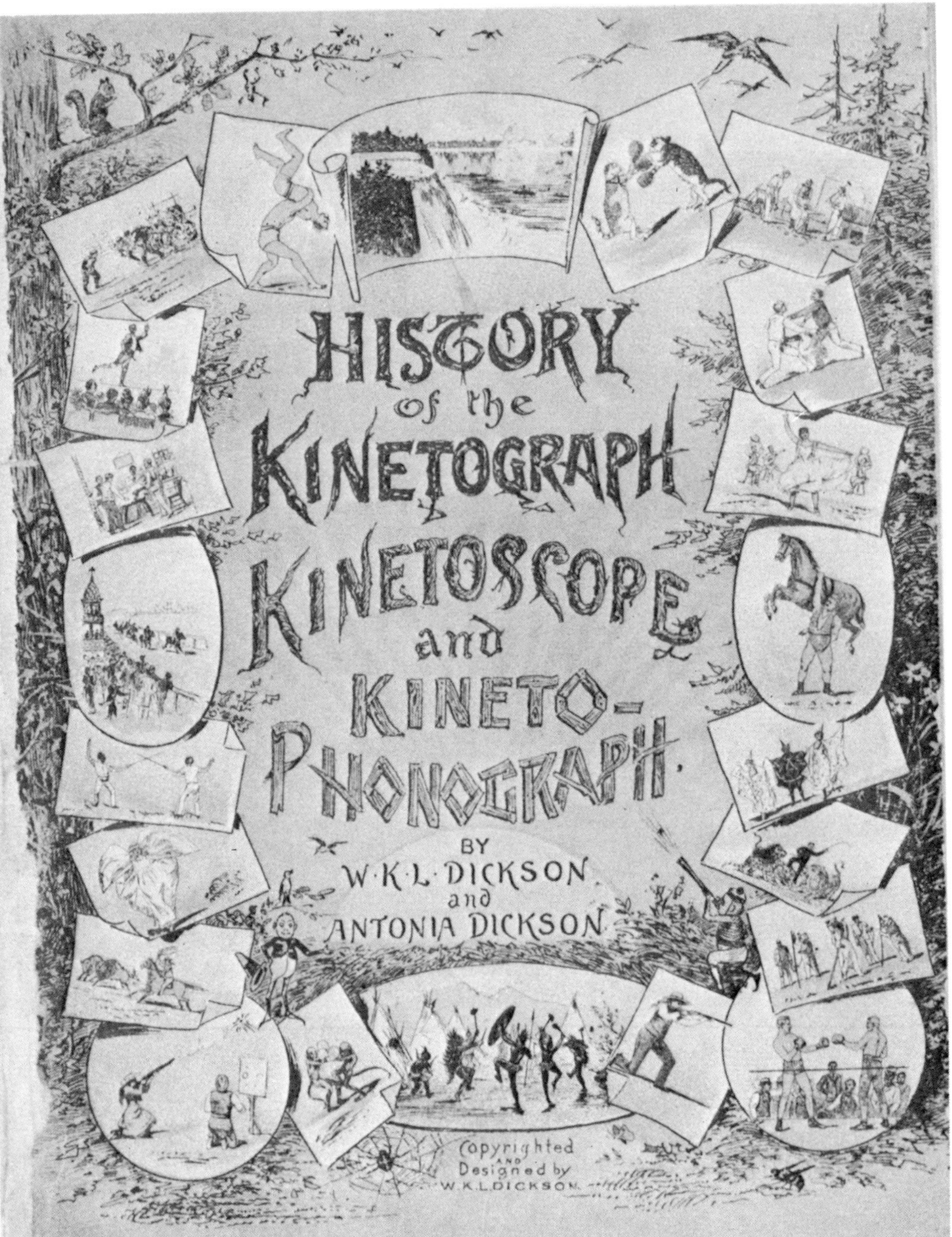
HISTORY
of the
KINETOGRAPH
KINETOSCOPE
and
KINETO-
PHONOGRAPH.
BY
W·K·L· DICKSON
and
ANTONIA DICKSON
Copyrighted
AND
Designed by
W.K.L. DICKSON

ALBERT BUNN,
IMPRIMATUR,
93 WORLD BUILDING, NEW YORK.

THOMAS A. EDISON.

In the year 1887, the idea occurred to me that it was possible to devise an instrument which should do for the eye what the phonograph does for the ear, and that by a combination of the two all motion and sound could be recorded and reproduced simultaneously. This idea, the germ of which came from the little toy called the Zoetrope, and the work of Muybridge, Marié, and others has now been accomplished, so that every change of facial expression can be recorded and reproduced life size. The Kinetoscope is only a small model illustrating the present stage of progress but with each succeeding month new possibilities are brought into view. I believe that in coming years by my own work and that of Dickson, Muybridge Marié and others who will doubtlessly enter the field that grand opera can be given at the Metropolitan Opera House at New York without any material change from the original, and with artists and musicians long since dead.
The following article which gives an able and reliable account of the invention has my entire endorsation. The authors are peculiarly well qualified for their task from a literary standpoint and the exceptional opportunities which Mr Dickson has had in the fruition of the work.

Thomas A Edison

THE KINETOGRAPHIC THEATRE AT EDISON'S LABORATORY.

HISTORY OF THE INVENTION.

IN the year 1887, Mr. Edison found himself in possession of one of those breathing spells which relieve the tension of inventive thought. The great issues of electricity were satisfactorily under way. The incandescent light had received its finishing touches; telephonic and telegraphic devices were substantially interwoven with the fabric of international life; the phonograph was established upon what seemed to be a solid financial and social basis, and the inventor felt at liberty to indulge in a few secondary flights of fancy. It was then that he was struck by the idea of reproducing to the eye the effect of motion by means of a swift and graded succession of pictures and of linking these photographic impressions with the phonograph in one combination so as to complete to both senses synchronously the record of a given scene. At the time of which we speak the conditions of photography were eminently favorable as a basis for experiments, while their obvious limitations afforded a tempting field for further research. The initial crudities of Daguerre, Niepce and their peers had been successfully tested and superseded, and the science was now enriched by the discoveries of Maddox, author of the bromo-gelatine process of instantaneous photography. The initial principle of moving images was suggested by a toy, familiar to children as the zoetrope, or wheel of life, a contrivance consisting of a cylinder some ten inches wide, open at the top, around the lower and interior rim of which a series of related pictures is placed, representing any given phase of life, susceptible of swift and continuous motion. The cylinder is then rapidly rotated, and the eye of the spectator, being directed to the narrow and vertical slits on the outer surface of the cylinder, is edified by a series of painfully spasmodic jerks which by the exercise of a liberal fancy may pass as dimly suggestive of human and equine antics. This effect, as the reader probably knows, springs from the substitution of one succeeding phase of an attitude for another so as to produce the effect of continuity upon the retina, and the failure of these successive images to blend into an illusive picture is due to the roughness of the woodcuts and the

INTERIOR VIEW PHOTOGRAPHIC STUDIO.

impossibility of obtaining the requisite degree of speed. The idea, however, was a novel and ingenious one, and as such commended itself to the attention of photographic experts such as Mr. Muybridge and others, who, working upon the delicately responsive surface of the Maddox dry gelatine plate, were able to secure greatly superior results. Despite these important improvements, however, the system presented grave imperfections, and the limited speed attainable militated hopelessly against the desired realism of effect. It was then that a series of experiments was entered upon at the Orange Laboratory, extending over a period of six years.

The synchronous attachment of photography with the phonograph was early contemplated in order to record and give back the impressions to the eye as well as to the ear. The comprehensive term for this invention is the kineto-phonograph. The dual taking machine is the phono-kinetograph, and the reproducing machine is the phono-kinetoscope, in contradistinction to the kinetograph and kinetoscope, which apply respectively to the taking and reproduction of movable but *soundless* objects.

The initial experiments took the form of microscopic pinpoint photographs, placed on a cylindrical shell, corresponding in size to the ordinary phonograph cylinder. These two cylinders were then placed side by side on a shaft, and the sound record was taken as near as possible synchronously with the photographic image, impressed on the sensitive surface of the shell. The photographic portion of the undertaking was seriously hampered by the materials at hand, which, however excellent in themselves, offered no substance sufficiently sensitive. How to secure clear-cut outlines, or indeed any outlines at all, together with phenomenal speed, was the problem which puzzled the experimenters. The Daguerre, albumen and kindred processes met the first requirements, but failed when subjected to the latter test. These methods were therefore regretfully abandoned, a certain precipitate of knowledge being retained, and a bold leap was made to the Maddox gelatine bromide of silver emulsion, with which the cylinders were coated. This process gave rise to a new and serious difficulty. The bromide of silver haloids, held in suspension with the emulsion, showed themselves in an exaggerated

PHOTOGRAPHY EXTRAORDINARY.

coarseness when it became a question of enlarging the pin-point photographs to the dignity of one-eighth of an inch, projecting them upon a screen, or viewing them through a binocular microscope. Each accession of size augmented the difficulty, and it was resolved to abandon that line of experiment and revolutionize the whole nature of the proceedings by discarding these small photographs and substituting a series of very much larger impressions, affixed to the outer edge of a swiftly rotating wheel or disk and supplied with a number of pins, so arranged as to project under the centre of each picture. On the rear of the disk, upon a stand, a Geissler tube was placed, connected with an induction coil, the primary wire of which, operated by the pins, produced a rupture of the primary current, which, in its turn, through the medium of the secondary current, lighted up the Geissler tube at the precise moment when a picture crossed its range of view. This electrical discharge was performed in such an inap-

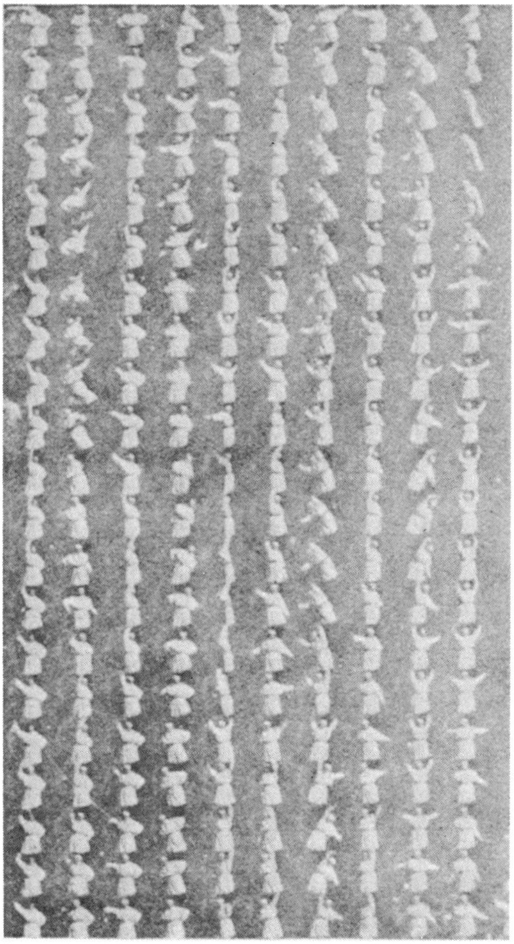
W. K. L. D

AN EARLY KINETOGRAPHIC EXPERIMENT.

A BEAR PARTY

preciable fraction of time, the succession of pictures was so rapid and the whole mechanism so nearly perfect that the goal of the inventor seemed almost reached. "We needs must love the highest," however, an axiom which holds good in science as well as in character, and the methods still pointed to possible improvement.

"TEDDY"

Then followed some experiments with drums, over which sheets of sensitized celluloid film were drawn, the edges being pressed into a narrow slot in the surface, similar in construction to the old tinfoil phonograph. A starting and stopping device was also applied, identical with the one used and in a later experiment explained in these pages. The pictures were then taken spirally to the number of two hundred or so, but were limited in size, owing to the rotundity of surface, which only brought the centre of the picture into focus. The sheet of celluloid was then developed, etc., and placed upon a transparent drum, bristling at its outer edge with brass pins. When the drum was rapidly turned these came in contact with the primary current of an induction coil, and each image was lighted up in the same manner as described in the previous disk ex-

A GENTLEMAN IN WAITING. W. K. L. D

periment, with this difference only, that the *inside* of the drum was illumined.

W. K. L. D.

The next step was the adoption of a highly sensitized strip of celluloid, one-half inch wide, but this proving unsatisfactory, owing to inadequate size, one-inch pictures were substituted on a band one and a half inches wide, the additional width being required for the perforations on the outer edge. These perforations occur at close and regular intervals, in order to enable the teeth of a locking device to hold the film steady for nine-tenths of the one-forty-sixth part of a second, when a shutter opens rapidly and admits a beam of light, causing an image or phase in the movement of the subject. The film is then jerked forward in the remaining one-tenth of the one-forty-sixth part of a second, and held at rest while the shutter has again made its round, admitting another circle of light, and so on until forty-six impressions are taken a second, or two thousand seven hundred and sixty a minute. This speed yields one hundred and sixty-five thousand six hundred pictures in an hour, an amount amply sufficient, when revolved before the eye, for an entertainment. In this connection it is interesting to note that were the spasmodic motions added up by themselves, exclusive of arrests, on the same principle that a train record is computed independently of stoppages, the incredible speed of twenty-six miles an hour would be shown.

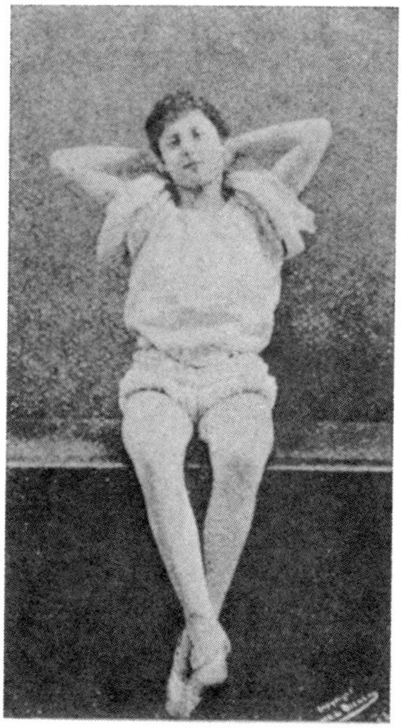

BERTOLDI.

The advantage of this system over a continuous band, and of a slotted shutter forging widely ahead of the film would be this, that only the fractional degree of light comprised in the $\frac{1}{2720}$ part of a second is allowed to penetrate to the film, at a complete sacrifice of all detail, whereas in the present system of stopping and starting each picture gets one-hundredth part of a second's exposure, with a lens but slightly stopped down, time amply sufficient, as any photographer knows, for the attainment of excellent detail, even in an ordinarily good light. Pre-eminently was this the case in using Messrs. Bausch and Lomb, and the Gundlach lenses, in which the opticians spared no skill or trouble in carrying out our specifications. It must be understood that but one camera is used for taking these strips, and not a battery of cameras. The next step after making the negative band is to form a positive or finished series of reproductions from the negative, which is passed through a machine for the purpose, in conjunction with a blank strip of film, which, after development and general treatment, is replaced in the kinetoscope or phono-kinetoscope, as the case may be. When a phonograph record has been taken simultaneously with such a strip, the two are started

BERTOLDI.

together by the use of a simple but effective device and kept so all through, the phonographic record being in perfect accord with the strip. In this conjunction, the tiny holes with which the edge of the celluloid film is perforated correspond exactly with the phonographic records, and the several devices of the camera, such as the shifting of the film and the operations of the shutter, are so regulated as to unify them with the different minutiæ of sound recording, one motor serving as a source of common energy to camera and phonograph, when they are electrically and mechanically linked together.

The establishment of harmonious relations between kinetograph and phonograph was a harrowing task and would have broken the spirit of inventors less inured to hardship and discouragement than Edison's veterans. The principles of this

PHOTOGRAPHY EXTRAORDINARY.
W. K. L. DICKSON. TAKEN BY HIMSELF.

doughty band, however, are based upon the sentiments of the indomitable Frenchman, who, in acceding to his lady-love's deprecatory request, remarked, " Madame, if the thing were difficult it would already have been done; if it is impossible it SHALL be done." The experiments have borne their legitimate fruit, and the most scrupulous nicety of adjustment has been achieved, with the resultant effects of realistic life, audibly and visually expressed.

The process of "taking" is variously performed, by artificial light in the photographic department, or by daylight under the improved conditions of the new theatre, of which we shall speak. The actors are kept as compact as possible and exposed either to the untempered glare of the sun, to the blinding effulgence of four parabolic manganese lamps, or to the light of twenty arc lamps, provided with highly actinic carbons, supplied with powerful reflectors equal to about 50,000 candle power. This radiance is concentrated upon the performers, while the kinetograph and phonograph are hard at work, storing up records and impressions for future reproduction.

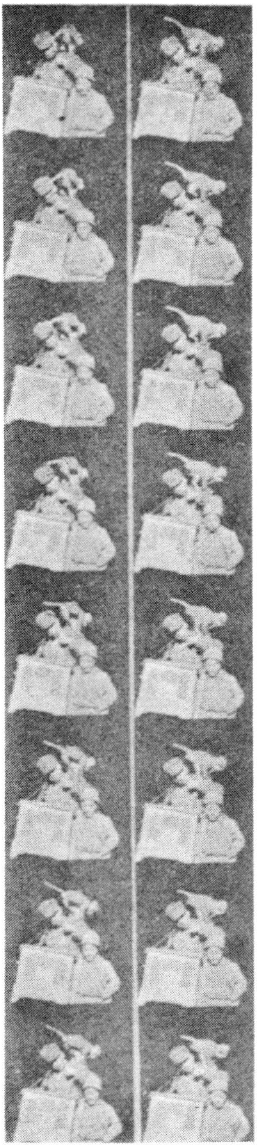

A popular and inexpensive adaptation of kinetoscopic methods is in the form of the well-known nickel-in-the-slot, a machine consisting of a cabinet, containing an electrical motor and batteries for operating the mechanism, acting as the impelling power to the film, which is in the shape of an endless band, fifty feet in length, the latter being passed through the field of a magnifying glass, perpendicularly placed. The photographic impressions pass before the eye at the rate of forty-six per second, through the medium of a rotating slotted disk, the slot exposing a picture at each revolution and separating the fractional gradations of pose. Projected against a screen on the kinetographic principle of stopping and starting or viewed through a magnifying glass, the pictures are eminently satisfactory, for the reason that the enlargement need not be more than ten times the original size. The projecting room, which is situated on the upper story of the photographic department, is hung with portentous black on exhibition evenings, in order to prevent any reflection from the circle of light emanating from the screen at the other end, the projector being placed behind a curtain of the same cheerful hue and provided with a single peephole for the accommodation of the lens. The effect of these sombre draperies and the weird accompanying monotone of the electric motor, attached to the projector,

E. J. MEEKER

INTERIOR OF THE KINETOGRAPHIC THEATRE, SHOWING KINETO-PHONOGRAPH.

are horribly impressive, and one's sense of the supernatural is heightened when a figure suddenly springs into his path, acting and talking with a vigor which leaves him totally unprepared for its mysterious vanishing. Projected stereoscopically, the results are

W. K. L. D

CARMENCITA THE SPANISH QUEEN OF DANCERS.

W. K. L. D.

"GAIETY GIRLS."

even more realistic, as those acquainted with that class of phenomena may imagine, and a pleasing rotundity is apparent which in ordinary photographic displays is conspicuous by its absence.

Nothing more marvelous or more natural could be imagined than these breathing, audible forms, with their tricks of familiar gesture and speech. The inconceivable swiftness of the photographic successions and the exquisite synchronism of the phonographic attachment have removed the last trace of automatic action, and the illusion is complete. The organ-grinder's monkey jumps upon his shoulder to the accompaniment of a strain of Norma. The rich tones of a tenor or soprano are heard, set in their appropriate dramatic action; the blacksmith is seen swinging his ponderous hammer exactly as in life, and the clang of the anvil keeps pace with his symmetrical movements; along with the rhythmical measures of the dancer go her soft-sounding footfalls; the wrestlers and fencers ply their intricate game, guarding, parrying, attacking, thrusting and throwing, while the quick flash of the eye, the tension of the mouth, the dilated nostril and the strong, deep breathing give evidence of the potentialities within. The crowning point of realism was attained on the occasion of Mr. Edison's return from the Paris Exposition of 1889, when Mr. Dickson himself stepped out on the screen, raised his hat and smiled, while uttering the words of greeting, "Good morning, Mr. Edison, glad to see you back. I hope you are satisfied with the kineto-phonograph."

The photographic rooms, with their singular completeness of appointment, have been the birth-place and nursery of the kinetoscope; and the more important processes connected with the preparation and development of the film, together with innumerable other mechanical and scientific devices, are still carried on in this department. The exigencies of natural lighting, however, incident to the better "taking" of the subjects and the lack of a suitable theatrical stage, necessitated the construction of a special building, which stands in the centre of that cluster of auxiliary houses which forms the suburbs of the laboratory, and which is of so peculiar an appearance as to challenge the attention of the most superficial observer. It obeys no architectural rules, it embraces no conventional materials and follows no accepted scheme of color. Its shape, if anything so eccentric can be entitled to that appellation, is an irregular oblong, rising abruptly in the centre, at which point a movable roof is attached, which is easily raised or lowered at the will of a single manipulator. Its color is a grim and forbidding black, enlivened by the dull lustre of myriads of metallic points; its materials are paper, covered with pitch and profusely studded with tin nails. With its great flapping sail-like roof and ebon complexion, it has a weird and semi-nautical appearance, like the unwieldy hulk of a

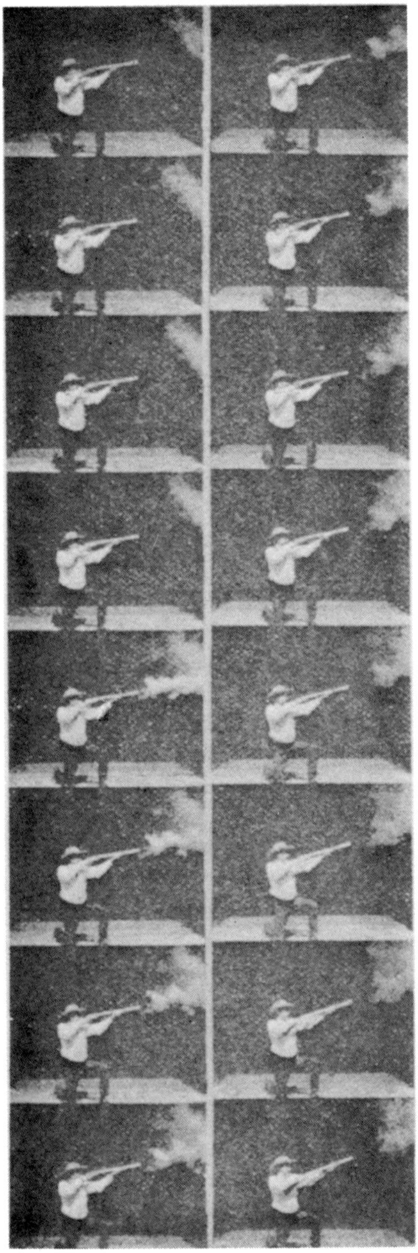

BUFFALO BILL. W. K. L. D.

medieval pirate-craft or the air-ship of some swart Afrite, and the uncanny effect is not lessened, when, at an imperceptible signal, the great building swings slowly around upon a graphited centre, presenting any given angle to the rays of the sun and rendering the apparatus independent of diurnal variations. The movable principle of this building is identical with that of our river swinging bridges, the ends being suspended by iron rods from raised centre-posts. This remarkable structure is known as the Kinetographic Theatre, or the "Black Maria," according to the mental cast of its sponsors. Entering, we are confronted by a system of lights and shades so sharply differentiated as to pain the eye, accustomed to the uniform radiance of the outer air.

As we peer into the illusive depths we seem transported to one of those cheerful banqueting halls of old, where the feudal chief made merry with human terrors, draping the walls with portentous black, and

INDIAN WAR COUNCIL.

thoughtfully providing a set of coffins for the accommodation of his guests. And what is this mysterious cell at the other extremity, sharply outlined against the dazzling radiance of the middle ground and steeped in an angry crimson hue? Are these inquisitorial dungeons, and is that lurid glare the advance guard of the awful Question? Is that gentle persuasive in process of administration, and do these half-guessed recesses conceal the hellish paraphernalia of rack and screw, glowing iron and crushing stone? Has the doom of ages overtaken our wizard at last, and is he expiating, with twisted limb and scorching flesh, the treasures of his unlawful wisdom? Ah, me! that the prosaic truth must be told. No dungeons are these, thrilling with awful possibilities, but simply a building for the better "taking" of kinetographic subjects. On the platform stand the wrestlers, pantomimists, dancers and jugglers, whose motions it is destined to immortalize. Against the nether gloom their figures stand out with the sharp contrast of alabaster basso-relievos on an ebony ground, furnishing a satisfactory explanation for the singular distinctness of the kinetographic strips. The lurid cell at the other end resolves itself into a compartment for changing the films from the dark box to the camera, the apparatus being run backward over a track leading from the black tunnel at the rear of the stage to this room, after which the door is shut and the films renewed for a fresh subject.

We have been sensible for some time of a disturbance of the ground beneath our feet, and are now aware that the building is slowly and noiselessly rotating on an axis, bringing into our range of vision the glory of the sun-rays westering to their close. Again we are reminded of that indissoluble chain of ideas which links the past with the present, and into the commonplace of existing facts come memories of that chamber in the golden house of Nero so arranged that "by means of skilfully planned machinery it moved on its axis, thus following the motions of the heavens, so that the sun did not appear to change in position, but only to descend and ascend perpendicularly."

K. L. D.

SIOUX INDIAN GHOST DANCE.

No department of the wizard's domains is more fraught with perennial interest than this theatre; none are more interwoven with the laughter, the pathos, the genius and the dexterities of life. No earthly stage has ever gathered within its precincts a more incongruous crew of actors, since the days when gods and men and animals were on terms of social intimacy; when Orpheus poured his melting lays into the ears of the brute creation, and gentle Anthony of Padua lured the suffering beasts to the mouth of his desert cave. The *Dramatis Personæ* of this stage are recruited from every characteristic section of social, artistic and industrial life, and from every conceivable phase of animal existence within the scope of laboratory enterprise. Bucking bronchos, terriers and rats, accomplished dogs who turn somersaults and describe serpentine dances, trained lions, bears and monkeys are among the stars of this unique company. On one occasion, the platform was occupied by a wire cage, the sometime arena for certain gallinaceous conflicts. A duel between two aspiring and vindictive roosters took place and the films have registered the strut, the swagger and the general bravado of the feathered knights. Another day chronicled the engagement of a troup of trained bears and their leaders, Hungarians by nationality, to whom the scientific eccentricities of the laboratory furnish an inexhaustible mine of wealth. The theatre at such times might move

A JAPANESE BEAUTY.

"The inextinguishable laughter of the blessed gods,"

and peals of these tantalizing cachinations are borne to the envious ears of dwellers in distant settlements, grim genii of the dynamo and ore milling departments. The men are tractable enough, when they can be induced to collect their scattery faculties and concentrate their attention on the exigencies of the situation; but the bears are divided between surly discontent and a comfortable desire to follow the bent of their own inclinations.

It is only after much persuasion that they can be lifted into higher planes of thought and feeling and induced to subserve the interests of art. Once launched, however, their performances are inimitable, and nothing within the range of comic histrionics can approximate the humor of these uncouth gambols. One furry monster waddled up a telegraph pole, to the soliloquy of his own indignant growls; another settled himself comfortably in a deep armchair, with the air of a post-graduate in social science; a third rose solemnly on his hind legs and described the measures of some unclassifiable dance, to the weird strains of his keeper's music. Another blandly ignored the invitation of his two-footed ally and endeavored to divert that gentleman's mind by licking his swarthy face. Another, more warlike in tone than his companions, accepted his keeper's challenge and engaged with him in a wrestling match, struggling, hugging and rolling on the ground with a force and scientific accuracy which

LASSO THROWER. W. K. L. D

"LET ME LOOK."

would have insured his promotion to the international championship, were equal justice dealt to man and beast in this partial world.

Cats have figured very amusingly in the laboratory shows. We had a consignment of these sent to us, fresh from their triumphs in Barnum's circus, but their Orange début was delayed, not only on account of the weather, which was unfavorable, but by reason of certain ophthalmic troubles, induced by extensive clawing at each other's optics. We placed them in solitary confinement and doctored them with eye salve, an operation attended with piercing yells and much opposition, and by the time the sun saw fit to emerge from his blanket of clouds, the company was in a condition to appear in their several feats of jumping hoops, trundling toy coaches, boxing, riding bicycles, etc. The crowning feature of the performance was a boxing match between two vigorous Toms, the suspicious animation of which pointed to something more than scientific fervor, if one might judge by the scratching, spitting and extensive shedding of fur which prevailed. The showman assured us, however, that "Pete and Jack" were the best of friends in their private capacity and generally elected to travel in the same cage.

Organ-grinders and monkeys have contributed liberally to the kinetographic collection. It is estimated that the classic soil of New Jersey supports three of these itinerant musicians to each square mile, while denying its patronage to Seidl or Damrosch. It may, therefore, be inferred that artistic supplies, of this nature at least, are in no immediate danger of depletion. Every few days a native of sunny Italy may be seen describing his eccentric orbit in the direction of the Orange Laboratory, intoning his torturous strains and administering finishing touches to the decayed elegance of the monkey's attire. No perceptible chasm differentiates employer and employee. Both are draped in picturesque remnants, both reside on the frayed outskirts of society, both are the victims of insatiable curiosity. On one occasion the nickel and slot was on view, a machine which, claiming only to be a duodecimo edition of the kinetoscope, and designed to meet the popular requirements on a simple and inexpensive basis, is limited in its scope, and admits only of a single spectator at a time, who is supposed to glue his eyes to the narrow opening at the top. It so happened that master and monkey were simultaneously stricken with the desire to see, an impulse which brought their heads into sharp contact and led to much energetic cuffing and chattering. The poor little ape went to the wall, as

CORBETT AND COURTNEY BEFORE THE KINETOGRAPH.

is generally the case with poor relations, and the Italian regaled himself with a long contemplation of his charms, after which he vacated in favor of his slave, whose delight was unbounded at the spectacle of these diminutive doubles, performing the familiar round of the itinerant repertoire. It is on record that the tiny creature laughed, actually laughed, oblivious for a few enchanted seconds of unkind man, of sunless cellars, starvation and chastisement, and the tribute is accepted as one of the most gratifying in

W. K. L. D.

CORBETT AND COURTNEY FIGHT.

all the archives of the sated kinetoscope. Monkeydom has an inexhaustible fund of varied emotions, underlying the unfathomable antiquity, the measureless sadness of its exterior, and we cannot doubt that Mr. Garner's researches will bring characteristics to light, which will be invaluable in the dissipation of ennui. The most *blasé* and self-contained member of the "four hundred" could hardly have opposed an impassive front to the antics of these prehistoric babies, these prophetic epitomes of man. One tiny Simian fell into ecstasies of delight over his reflected image. He postured before it, felt it with his supple fingers, chattered with easy sociability, gesticulated, danced, growled and finally darted to the back of the mirror, impelled by a spirit of laudable investigation and a desire to determine the nature and origin of the phenomenon, repeating his efforts until well assured that the mystery lay beyond the scope of his mental horizon.

In this connection it may be said that Mr. Garner, the student of animal speech, has been among our constant visitors, and has followed the development of the kinetograph with intelligent attention. Mr. Garner's recent journey into central Africa, undertaken for the purpose of verifying his theory as to the existence of a Simian tongue, has yielded such gratifying results as to spur the explorer on to fresh endeavor. Entrenched within the recesses of an enormous cage, which was successively transported to the most teeming centres of animal life, and which was so constructed as totally to conceal the occupant, Mr. Garner was enabled to study our primitive ancestors at his ease, and to store his mind with a variety of impressions, humorous, pathetic and utilitarian.

He purposes a second trip shortly, and will bring the wide resources of the kinetograph to bear upon many additional phases of animal life, so that our aristocratic sybarites may enjoy the thrilling dramas of jungle and forest, without imperilling that "repose which is the essential attribute of good breeding," or embrowning the delicate cuticle on their patrician countenances. When music and

W. K. L. D.

ALCIDE CAPITAINE.

W. K. L. D.

MEXICAN KNIFE DUEL.

oratory and histrionics have lost their power to charm, they may ensconce themselves in the yielding recesses of ruby or violet velvet thrones, with the scent of hot-house flowers around and the memory of a Lucullus feast titillating the cultured palate. From that luxurious stronghold they may contemplate the awful rush of maddened brutes, the tawny flash of the savage eyes, the lightning play of the vigorous muscles, may hear the shock of the reverberating earth, the roar of the great cats, the grinding of fangs, the tearing of iron claws, the scream of the dying elephant, the sardonic laugh of the merciless hyena,—robber and violater of the dead,—the whirr of mighty pinions, as the vultures descend to their ghastly feast, all the "travailing and groaning" of burdened creation. The sunlight will tremble through the leafy arcades and cast its fantastic shadows on the opulent growth; will extract, through its own unapproachable alchemy, each superb *nuance* on tree and flower and creeper; will vivify the tawny beauty of tiger and lion, and give to the lustrous plumage of the bird an added glory; will burnish the scale of the Ophidian and encase the flashing Coleoptera in mail of metal and gem. All the kingdoms of the world, with their wealth of color, outline and sound, shall be brought into the elastic scope of individual requirement at the wave of a nineteenth-century wand.

MME ARMAND 'ARY.

Of human subjects we have a superfluity, although the utmost discrimination has been observed in the selection of themes. In point of classical beauty and as a prophetic exposition of what we may expect in the physical regeneration of the race, Eugen Sandow, the modern Hercules, stands foremost. From an anatomical point of view, this great athlete has attained ideal perfection of form, combined with phenomenal strength and grace. The gladiatorial records of Rome contain no evidence of muscular development such as Sandow presents. His normal chest development is forty-seven inches;

SANDOW.

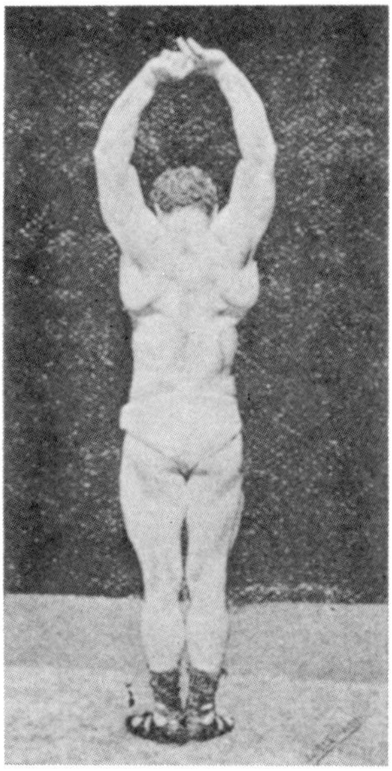

SANDOW.

expanded, it is sixty-one, showing an increase of fourteen inches. The greatest expansion ever known at the Olympic games was six. This is demonstrated in the kinetograph series, together with the more remarkable feats relating to the action and uses of the various muscles, such as the lifting of three-hundred pound dumb-bells at arm's length over his head, and the sustaining of a platform and three horses on his chest, a dead weight of over three thousand pounds. That his agility is equal to his strength is shown in the

Copyright FALK.

EUGEN SANDOW, THE MODERN HERCULES.

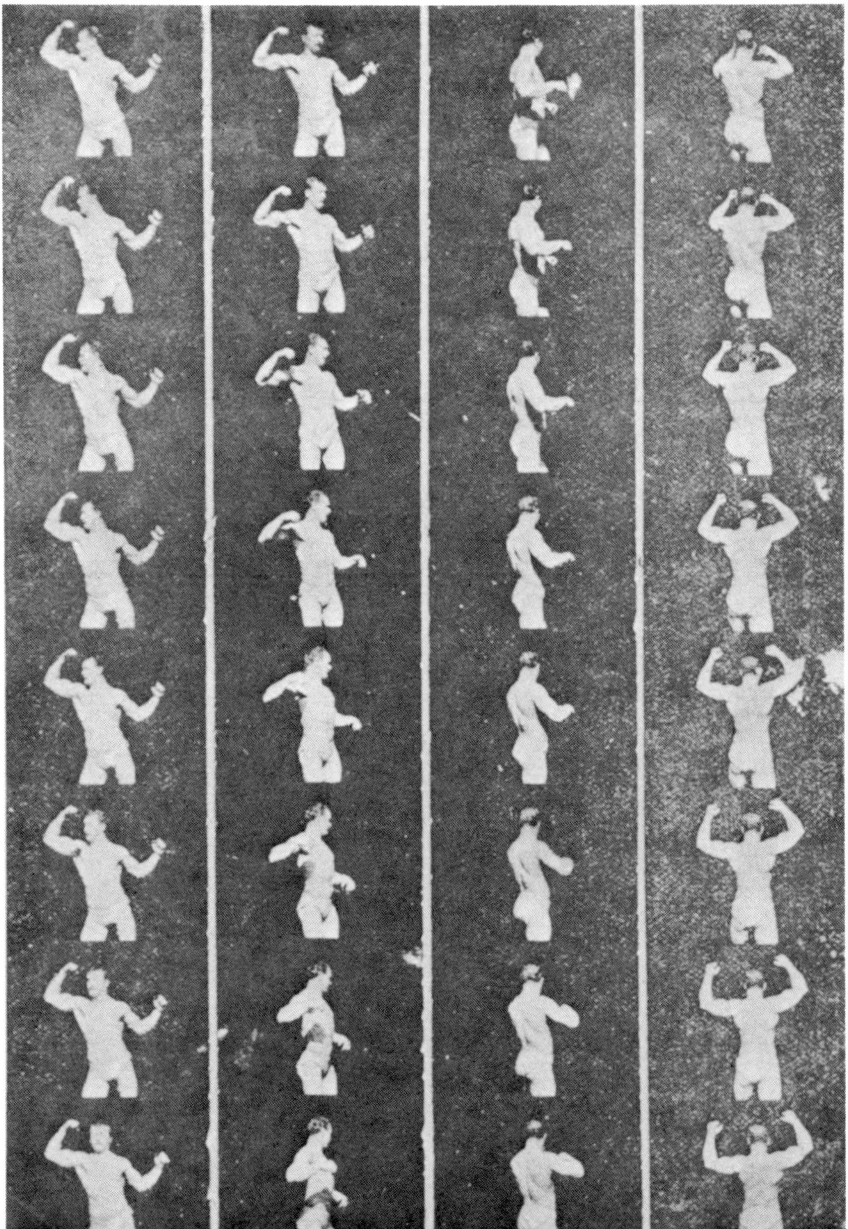

W. K. L. D.

EUGEN SANDOW.

W. K. L. D.

FENCERS.

exploit known as the Roman column, where, with knees chained to an iron column, he bends backwards to the ground, lifting huge dumb-bells and three men to show the use of the dorsal and abdominal muscles. Many of his feats and poses are modeled upon the pictured achievements of ancient bards and sculptors, and the effect is artistically perfect.

In the suggestion of unguessed planes of virility, Buffalo Bill and his motley suite have materially assisted. Nothing more stimulating can be imagined than these unconventional types of humanity, these riotous Texan cow-boys and Mexicans, these Moors, Arabs and Indians, riding, lassoing, shooting, juggling and sparring with the swiftness and ease born of untrammeled physical conditions. Unique in interest also is the Omaha war dance, the Sioux ghost dance and Indian war council, features of aboriginal life which may be historically valuable long after our polished continent has parted with the last traces of her romantic past.

Each day is fruitful in kinetographic developments. The records embrace the demonstration of pugilism by Corbett—the most refined and skilful exponent of

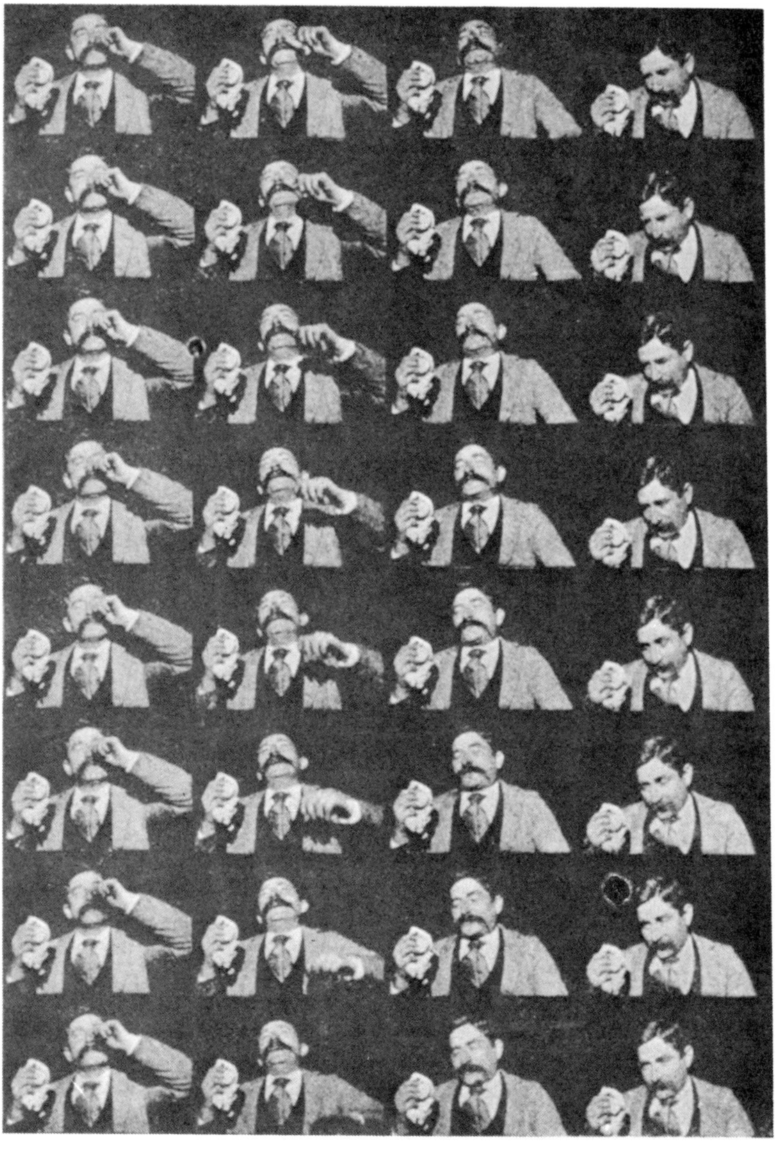

W. K. L. D.

RECORD OF A SNEEZE.

the art of self-defence; eastern gun-juggling and knife-tumbling by Sheik Hadji Tahar—an illustrious Moor, high in favor with the Sultan; Mexican duelling by Pedro Esquivel and Dionecio Gonzalez; the exploits of Vicente Ore Passo, the champion lasso thrower; the pugilistic and boxing parodies of Glenroy brothers and of Walton and Slavin. The broadsword exercises of the Englehart sisters; the antics of the pickaninnies from the "Passing Show"; the dancing of the three Sarashe sisters; the sword combats and acrobatism of Salem Nassar and Nazib; the whirlwind gun-spinning of Kessell; the Japanese twirling of Toyo Kichu; the marvelous sharp-shooting of Buffalo Bill and Annie Oakley; the perilous feats of Caicedo, king of the wire, and of Mme. Alcide Capitaine, on the trapeze; the terpsichorean measures of Carmencita, and the houris yclept "Gaiety-girls" from the Prince of Wales' Theatre, London, and a thousand others which it would be impossible to chronicle specifically.

JOHN WILSON.

Among these unclassifiable subjects are our recent mirthful experiences in securing the record of a sneeze. The victim was requested to assume a seat and favor the audience with that mild convulsion, and to the furtherance of that end, a large pinch of snuff was administered, the operator standing meanwhile in readiness, so as to catch the results in a graded series of one thousand pictures. A breathless silence ensued, the victim's face screwed, puckered and collapsed. There was evidently a hitch somewhere in the anatomical machinery. A second and larger pinch was administered, with no better result; a dose of ground tobacco followed, capped by a generous portion of black pepper. In vain. The wretched youth coughed, choked, sniffed, finally dissolving into tears, and amid shouts of laughter the attempt was abandoned, only to be renewed a few days later when the desired results were secured. Science hath her martyrs as well as religion.

Our quiver is almost full, and we have not touched as yet upon the microscopic subjects, a class of especial interest as lying

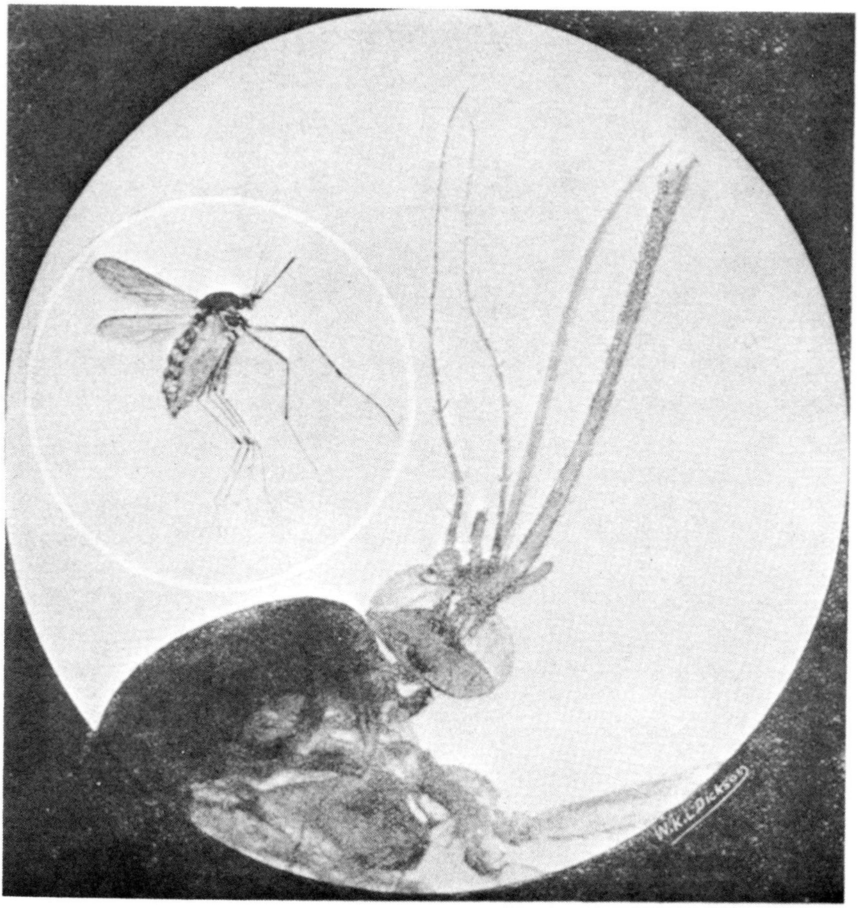

MICROSCOPIC ENLARGEMENT OF A MOSQUITO, SHOWING FEELERS, LANCE AND SUCTION TUBE.

outside of the unaided vision of man. In the treatment of these infinitesimal types, much difficulty was experienced in obtaining a perfect adjustment so as to reproduce the breathing of insects, the circulation of blood in a frog's leg, and other attenuated processes of nature. The enlargement of animalculæ in a drop of stagnant water proved a most exacting task, but by the aid of a powerful lime-light concentrated on the water, by the interposition of alum cells for the interception of most of the heat rays, and by

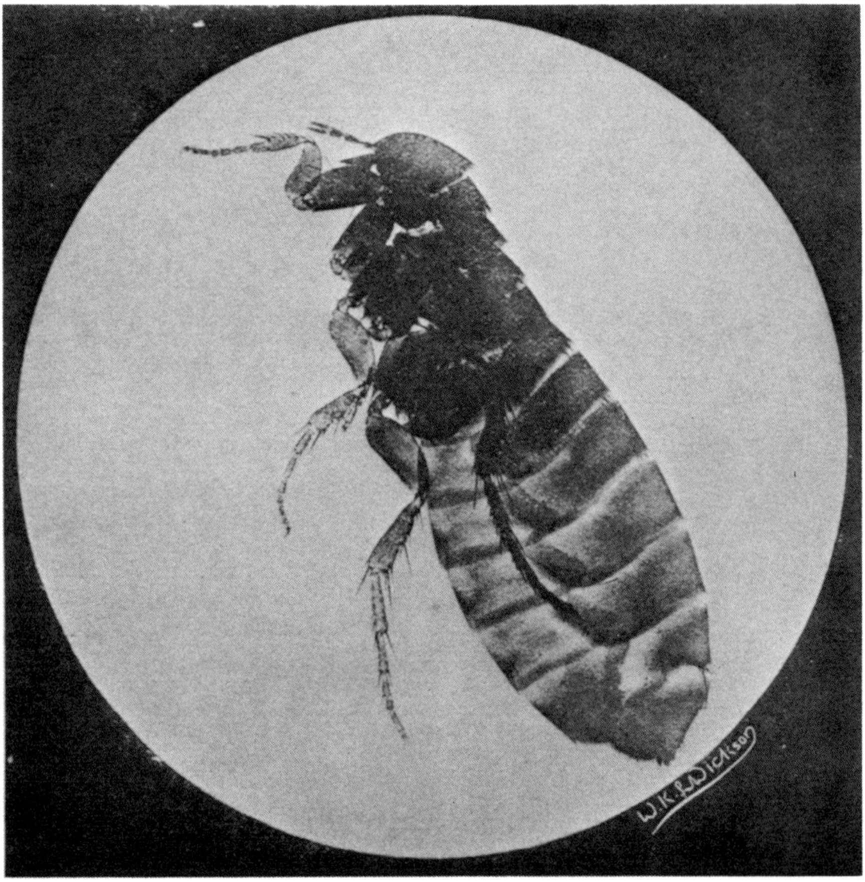

MICROSCOPIC ENLARGEMENT OF A FLEA.

the use of a quick shutter and kindred contrivances, the obstacles were overcome and the final results were such as fully to compensate for the expenditure of time and trouble. We will suppose that the operator has at last been successful in imprisoning

W. K. L. D.
BOXING CATS.

tricksy water-goblins on the sensitive film, and in developing the positive strip and placing it in the projector. A series of inch-large shapes then springs into view, magnified stereoptically to nearly three feet each, gruesome beyond power of expression, and exhibiting an indescribable celerity and rage. Monsters close upon each other in a blind and indiscriminate attack, limbs are dismembered, gory globules are tapped, whole batallions disappear from view. Before the ruthless completeness of these martial tactics the Kilkenny cats fade into insignificance, and the malign Jersey mosquito resolves itself into an honorable champion, sounding the bugle of approach and defiance. A curious feature of the performance is the passing of these creatures in and out of focus, appearing sometimes as huge and distorted shadows, then springing into the reality of their own size and proportions. Investigations in this line, while enriching the general mental store, are scarcely conducive to domestic comfort. An unseen enemy is usually voted to be peculiarly undesirable, but who would not close their eyes to the unimaginable horrors which micro-photography reveals in connection with the kinetoscope? Who would not prefer the mosquito as we know him, a brace of gossamer wings, a tiny bugler in the insectiferous ranks of creation, to this monstrous Afrite with its hungry and innumerable eyes, its ribbed and

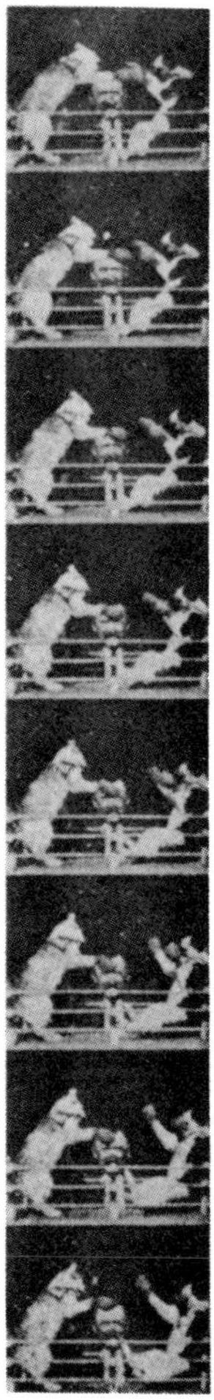

W. K. L. D.
BOXING CATS.

W. & L. D.

SHEIK HADJI TAHAR.

W. K. L. D.

WALTON AND SLAVIN—FROM "1492."

bat-like pinions, and its formidable arsenal of weapons? Who would not prefer the squat little speck which we know as the flea, and whom, if we do not exactly welcome, we at least tolerate, to this hideous apparition, with its bristling body and its lobster-like shape? What ratio does the comely and affectionate house-fly bear to this spiky argus-eyed demon of the microscope? Dead specimens are bad enough in all conscience, mere spreadings of animal matter as they are, but when the kinetoscope steps in and reveals the malign activities beneath these awful shapes, the merciless plunge of the bill and its cargo of human gore, the sharp stab of the long serrated lance, the hypodermic injections of virulent poison, then are we indeed in evil case and disposed to murmur dubiously, "Where ignorance is bliss 'tis folly to be wise."

Thought is the great, indeed the only constructive material which we possess; the imagination or image-forming faculty is immeasurably potent in the moulding

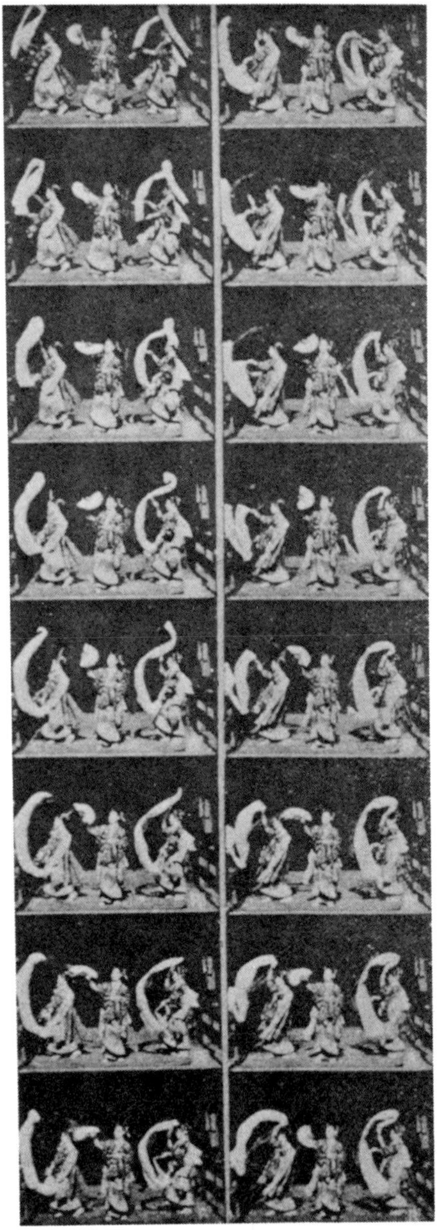

W. K. L. D.

JAPANESE DANCERS.

of mind and body. What will be the effect of this torrent of appalling impressions upon the mental and physical tissues? Imagine a super-sensitive brain transported from these enlarged monstrosities to the magnifying lenses of dreamland. What howls of mortal anguish one may expect and what an unpleasant stir generally in that gentleman's domestic economy. Then, inasmuch as

"of the soule the bodie forme doth take
For soule is forme and dothe the bodie make,"

our globe is likely to be peopled by singular modifications of

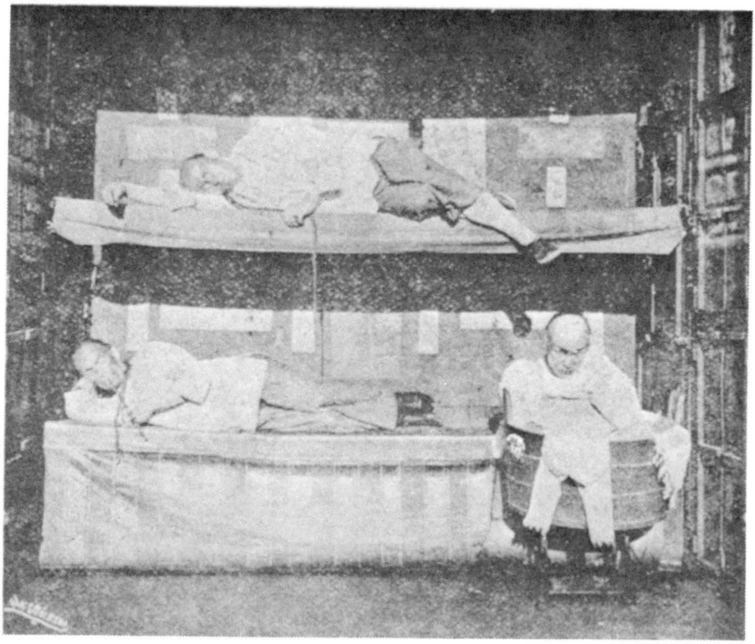

Kinetoscopic Image four times enlarged.
OPIUM DEN.

existing types, engendered by the frightful scientific discoveries of the day, so that we may confidently look forward to a race of beings before which the twin horrors of Sicily, the Chimeras, the Cyclops and Centaurs, the triple-headed dogs and seven-headed serpents of ancient Greece are respectable and humdrum characters.

Hitherto we have limited ourselves to the delineation of detached subjects, but we shall now devote some space to one of our most ambitious schemes, of which these scattered impersonations are but the heralds. Preparations have long been on foot to extend the number of the actors and to increase the stage facilities, with a view to the presentation of an entire play, set in its appropriate frame. Although necessarily crude, these efforts undoubtedly contain the germs of ultimate success. We have

Kinetoscopic Image four times enlarged.

ROW IN A CHINESE LAUNDRY.

enacted quite a number of experiences, one of which is supposed to take place in a barber's shop. The establishment is fitted up with a chair, shaving apparatus and everything suited to the requirements of tonsorial art, while a flaring sign invites the "Great Unshorn" to relieve themselves of their superfluous hirsute adornments at the modest expenditure of a nickel. The customers enter—rather a shady lot, as may be inferred from the

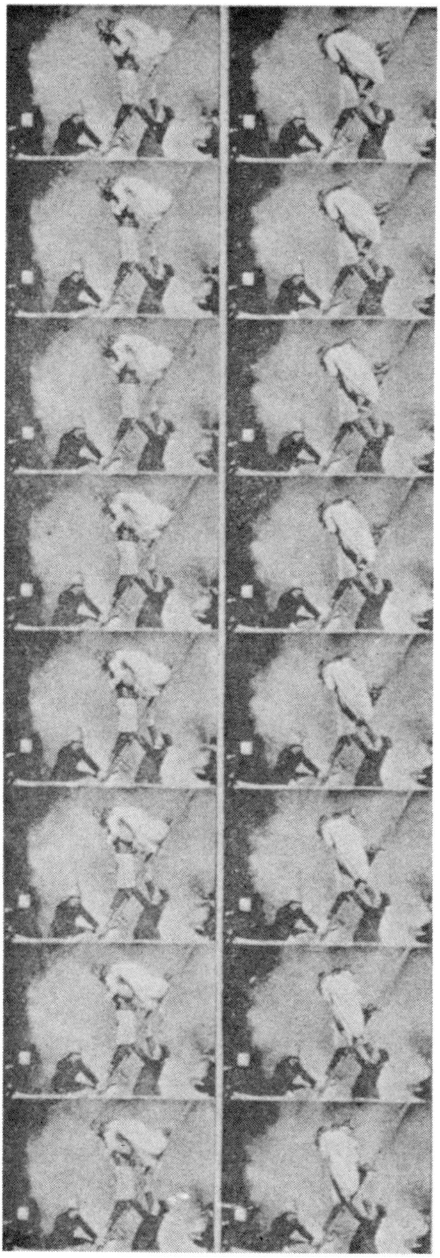

FIRE RESCUE. W. K. L. D.

prices,—are received by the artist (a brilliant young Greek professional who kindly lent his services), are hustled unceremoniously into the chair, lathered, scraped and despoiled of their locks in business-like fashion and with small regard for tender susceptibilities. The operation over, "Next" is roared out imperiously, and the summons are responded to by a stampede of excited competitors. A scrimmage follows, in the effort to secure the solitary chair, the barber attempts in vain to restore order, and the scene ends with a general fight.

On another occasion, an impromptu picnic was rigged up just outside the stately grounds of the "Black Maria." The actors fell readily into position, being possessed of a good deal of unsuspected histrionic talent, and the scene was soon animated in the extreme. A fire was lit, a gipsy kettle hung, and some one commenced energetically to blow the embers; crockery and comestibles were unpacked and set out invitingly by several deft-handed maidens, while a *mauvais sujet*, oblivious of the presence of these charmers, vaulted upon the table, drew a beaker of foaming ale and started a general frolic.

A highly condensed *finale* from Hoyt's "Milk White Flag" was also given, comprising twenty-four actors, together with certain thrilling Mongolian experiences, enacted respectively in a laundry and opium den. These are quite lively, necessitating, as they do, the intervention of the police, and an indiscriminate melée. Possibly the most exciting scene in our repertoire is a fire rescue with the stage-honored accessories of ladders, a burning house, clouds of steam and smoke, and a lovely female, airily clad, leaping into the extended arms of a gallant hero.

To the final development of the kinetographic stage, than which no more powerful factor for good exists, no limitations can possibly be affixed. The shadowy histrionics of the near future will yield nothing in realistic force and beauty to their material sisters. No imperfections will mar the illusion. The rich strain of a Seidl or Damrosch orchestra, issuing from a concealed phonograph, will herald the impending drama, and attune the hearts of the expectant throng. The curtain will rise, exposing some one of the innumerable phases of pictorial art, some soft English pastoral or cosy interior of a mansion, some fastness in the Alps or Himalayas, some tempestuous ocean scene, quickened with the turbulent

"MILK WHITE FLAG."

anguish of the unresting sea, some exquisite landscape, steeped in the jeweled lights of sunset or the roseate sheen of morn. The actors will enter singly and in groups, in the graceful interweaving of social life, the swirl of the dance or the changeful kaleidoscope of popular tumult. The tones will be instinct with melody, pathos, mirth, command, every subtle intonation which goes to make up the sum of vocalism; the clang of arms, the sharp discharge of artillery, the roll of thunder, the boom of ocean surges, the chant of the storm wind, the sound of Andalusian serenades and the triumphant burst of martial music,—all these effects of sight and sound will be embraced in the kinetoscopic drama, and yet of that living, breathing, moving throng, not one will be encased in a material frame. A company of ghosts, playing to spectral music. So may the luminous larvæ of the Elysian Fields have rehearsed earth's well-beloved scenes to the exiled senses of Pluto's Queen.

This line of thought may be indefinitely pursued with application to any given phase of outdoor or indoor life which it is desired to reproduce. Our methods work admirably and every day adds to the security and the celerity of the undertaking. No scene, however animated and extensive, but will eventually be within reproductive power. Martial evolutions, naval exercises, processions and countless kindred exhibitions will be recorded for the leisurely gratification of those who are debarred from attendance, or who desire to recall them. The invalid, the isolated country recluse, and the harassed business man can indulge in needed recreation, without undue expenditure, without fear of weather, without danger to raiment, elbows and toes, and without the sacrifice of health or important engagements. Not only our own resources but those of the entire world will be at our command, nay, we may even anticipate the time when sociable relations will be established between ourselves and the planetary system, and when the latest doings in Mars, Saturn and Venus will be recorded by enterprising kinetographic reporters.

The advantages to students and historians will be immeasurable. Instead of dry and misleading accounts, tinged with the exaggerations of the chroniclers' minds, our archives will be enriched by the vitalized pictures of great national scenes, instinct with all the glowing personalities which characterized them.

What is the future of the kinetograph? Ask rather, from what conceivable phase of the future it can be debarred. In the promotion of business interests, in the advancement of science, in the revelation of unguessed worlds, in its educational and re-creative powers, and in its ability to immortalize our fleeting but beloved associations, the kinetograph stands foremost among the creations of modern inventive genius. It is the crown and flower of nineteenth-century magic, the crystallization of Eons of groping enchantments. In its wholesome, sunny and accessible laws are possibilities undreamt of by the occult lore of the East; the conservative wisdom of Egypt, the jealous erudition of Babylon, the guarded mysteries of Delphic and Eleusinian shrines. It is the earnest of the coming age, when the great potentialities of life shall no longer be in the keeping of cloister and college, sword or money-bag, but shall overflow to the nethermost portions of the earth at the command of the humblest heir of the divine intelligence.

REPOSE.

FINIS.

Notes on facts re the pioneer days of the perforated Film moving photog. work at Edisons Lab. completed End of 1888 – including –

- Taking Camera – 40 pictures a second. Size picture settled & designed by Experiment
- as 3/4" x 1/2" <u>As now in use</u> –
- perforated top & bottom 4 to the pict. <u>As now</u>
- Sprocket wheel to fit perforations. " "

Shutter but slightly altered <u>As now in use</u>

<u>Projecting</u> – with the same (Taking camera) machine – the sprocket wheel only exchanged for one slightly smaller to allow for shrinkage of ~~film~~ Positive (name given <u>by</u> Edison) after developement – perforations to fit a slightly smaller sprocket wheel –

Printing by contact – neg. over virgin pos. films fed over drum running continuously –

Developing on large drums –

The above machine we called the Kinetograph – later in 1889 was <u>synchronized</u> with the phonograph & projected on the screen – See Journals & witness.

Yet there are many who claim precedence, though actually coming in after the work was done – on the perforated film – Stop motion & machine were created bet. 1887 & end of 1888–9

Geo Eastman made the first film <u>Exclusively</u> for us, – so lets give Edison full credit for this work.

W. K. Laurie Dickson

If the general public were asked to name the most important photographic discovery of the year, there is little doubt but what they would be unanimous in declaring the kinetoscope. Although the idea is by no means a new one, Muybridge, Anschütz, Marey and others have already worked at the same theory on a comparatively small scale, yet the kinetograph can in no way be compared with the machines of other workers. While the latter consisted of a battery of at most twenty-four cameras, exposed by electricity, the kinetograph is but a single camera, capable of making many thousands of photographs in the space of a few seconds. When Edison conceived the idea the working out of the arrangement was intrusted to Mr. W. K. L. Dickson, a clever young electrical engineer. There are few who really can imagine the thousands of difficulties that required to be overcome before such an apparatus could be made to work satisfactorily. Imagine a roll of film one hundred and fifty feet long, which, in the space of 90 seconds, had to have impressed upon it nearly three thousand images. Imagine the delicate and perfect arrangement that had to be constructed to bring this film accurately into place, and expose it to the light each time without the slightest vibration—without the slightest hitch of any kind. Those who have studied the details of the arrangement pronounce it to be simply marvelous, and reflecting the highest credit on the clever engineer. Some account of this gentleman's career will, we are sure, be interesting to our readers. William Kennedy-Laurie Dickson was born in France and educated in England. His father, James Dickson, was a distinguished English painter and lithographer; many artists are numbered in his ancestral roll, among others the great Hogarth. His mother was Miss Elizabeth Kennedy-Laurie, of Woodhall, Kirkcudbright, Scotland, a brilliant scholar, musician, and renowned for her beauty, [redacted]. She was a descendant of the Lauries of Maxwellton, immortalized in the celebrated ballad, "Annie Laurie," and the Robertsons, of Strowan, connected with the Earl of Cassilis, the Duke of Athol, and the Royal Stuarts.

In his youth young Dickson gave evidence of a strong disposition to electrical experiments. Edison was his favorite scientific hero, and his youthful ambition

Publisher's Note:
All markings on the last two pages in the original edition are those of the author and have not been retouched for this edition by the request of the owner.

was not realized until, in 1879, he came to America, and, two years later, although but 21 years of age, attached himself to the great inventor in the capacity of superintendent of the testing and experimental department at the works in Goerck Street, New York. He was prominent in the development of the Electric Light Co., was chief electrician in the Edison Electric Tube Co., of Brooklyn, and was charged with the office of laying the first telegraphic and telephone wires underground in New York City during Mayor Grant's administration.

He is now chief of the electro-mining and kinetographic work at Edison's laboratory in Orange, having matured the magnetic separation of iron and other ores; is co-patentee with Edison of magnetic ore separators, and has completed for the inventor the kinetograph, kinetoscope, and phono-kinetoscope.

In this connection Mr. Edison, in the June number of *The Century*, thus speaks of Mr. Dickson:

"I believe that, in coming years, by my own work and that of Dickson, Muybridge, Marey, and others who will doubtless enter the field, that grand opera can be given at the Metropolitan Opera House at New York without any material change from the original, and with artists and musicians long since dead."

Messrs. Crowell & Co., of New York and Boston, whose specialty in classical *éditions de luxe* is universally recognized, are issuing a magnificently bound and illustrated biography of Edison, written by Mr. Dickson and his sister, Miss Antonia Dickson, which, in the opinion of leading judges, is as notable for wide scientific information as for purity of diction, dramatic force and lucidity of style. The peculiar advantages enjoyed by the authors in over thirteen years' intimacy with the inventor are shown in the accuracy and varied interest of the biography.

We give herewith some of the kinetograph pictures taken by Mr. Dickson, and a portrait of that gentleman.

Photography in colors still appears far away from our grasp. The improvements made by the Lumière Bros. upon the Lipmann process are of considerable interest, but it is very doubtful if this method of obtaining colors by interference is ever likely to result in a practical process.

Such have been, as well as we can call to mind, the principal discoveries and improvements during the past twelvemonth. What wonders will the coming year bring forth? Who can tell?

We have, fortunately, men—clever men—among us who are ever trying to invent and to improve. Men who devote their lives to the advancement of photography with but little reward, for who ever heard of a photographic inventor growing rich by his labors? But they work on, satisfied with their efforts to benefit their fellow creatures and the fascination that always exists in experimental work.

W. E. WOODBURY, Editor.

75 07212